Houghton
Mifflin
Harcourt

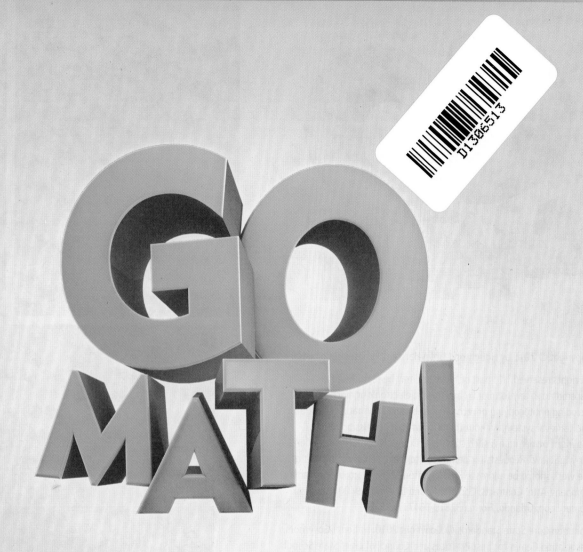

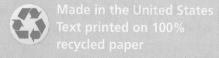

Houghton
Mifflin
Harcourt

Printed in the U.S.A.

ISBN 978-0-544-34232-3

19 0877 19

4500788160   C D E F G

Dear Students and Families,

Welcome to **Go Math!**, Grade 5! In this exciting mathematics program, there are hands-on activities to do and real-world problems to solve. Best of all, you will write your ideas and answers right in your book. In **Go Math!**, writing and drawing on the pages helps you think deeply about what you are learning, and you will really understand math!

By the way, all of the pages in your **Go Math!** book are made using recycled paper. We wanted you to know that you can Go Green with **Go Math!**

Sincerely,

The Authors

Made in the United States
Text printed on 100% recycled paper

GO MATH!

Authors

Juli K. Dixon, Ph.D.
Professor, Mathematics Education
University of Central Florida
Orlando, Florida

Edward B. Burger, Ph.D.
President, Southwestern University
Georgetown, Texas

Steven J. Leinwand
Principal Research Analyst
American Institutes for
    Research (AIR)
Washington, D.C.

Contributor

Rena Petrello
Professor, Mathematics
Moorpark College
Moorpark, California

Matthew R. Larson, Ph.D.
K-12 Curriculum Specialist for
    Mathematics
Lincoln Public Schools
Lincoln, Nebraska

Martha E. Sandoval-Martinez
Math Instructor
El Camino College
Torrance, California

English Language Learners Consultant

Elizabeth Jiménez
CEO, GEMAS Consulting
Professional Expert on English
    Learner Education
Bilingual Education and
    Dual Language
Pomona, California

Fluency with Whole Numbers and Decimals

  Critical Area Extending division to 2-digit divisors, integrating decimal fractions into the place value system and developing understanding of operations with decimals to hundredths, and developing fluency with whole number and decimal operations

2 Divide Whole Numbers   85

COMMON CORE STATE STANDARDS

5.NBT Number and Operations in Base Ten
Cluster B Perform operations with multi-digit whole numbers and with decimals to hundredths.
5.NBT.B.6

5.NF Number and Operations–Fractions
Cluster B Apply and extend previous understandings of multiplication and division to multiply and divide fractions.
5.NF.B.3

GO DIGITAL

Go online! Your math lessons are interactive. Use *i*Tools, Animated Math Models, the Multimedia eGlossary, and more.

Chapter 2 Overview

In this chapter, you will explore and discover answers to the following **Essential Questions**:

- How can you divide whole numbers?
- What strategies have you used to place the first digit in the quotient?
- How can you use estimation to help you divide?
- How do you know when to use division to solve a problem?

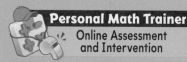
Personal Math Trainer
Online Assessment and Intervention

FOR MORE PRACTICE
GO TO THE
Personal Math Trainer

Practice and Homework

Lesson Check and
Spiral Review in
every lesson

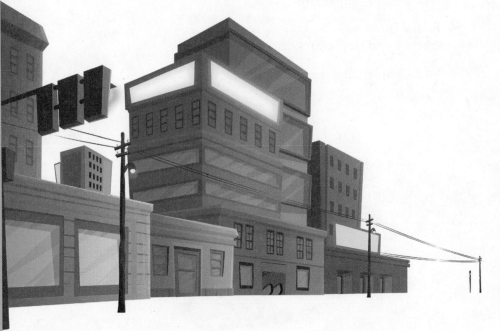

Divide Whole Numbers

Show What You Know

Personal Math Trainer
Online Assessment and Intervention

Check your understanding of important skills.

Name _____

▶ **Meaning of Division** **Use counters to solve.** (3.OA.A.2)

1. Divide 18 counters into 3 equal groups. How many counters are in each group?

_____ counters

2. Divide 21 counters into 7 equal groups. How many counters are in each group?

_____ counters

▶ **Multiply 3-Digit and 4-Digit Numbers** **Multiply.** (4.NBT.B.5)

| 3. | 321 × 4 | 4. | 518 × 7 | 5. | 4,092 ×   6 | 6. | 8,264 ×   9 |

▶ **Estimate with 1-Digit Divisors** **Estimate the quotient.** (4.NBT.B.6)

7. 2)312 8. 4)189 9. 6)603 10. 3)1,788

Math in the Real World

The height of the Gateway Arch shown on the Missouri quarter is 630 feet, or 7,560 inches. Find how many 4-inch stacks of quarters make up the height of the Gateway Arch. If there are 58 quarters in a 4-inch stack, how many quarters high is the arch?

Vocabulary Builder

▶ Visualize It ••••••••••••••••

Complete the Flow Map using the words with a ✓.

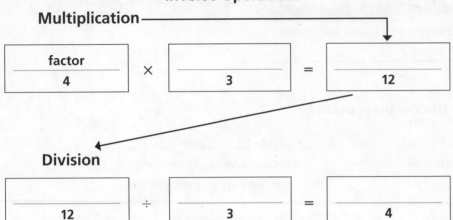

Inverse Operations

Multiplication

factor		
4		

×

3		

=

12		

Division

12		

÷

3		

=

4		

Review Words

compatible numbers
✓ dividend
✓ divisor
estimate
✓ factor
partial quotients
✓ product
✓ quotient
remainder

▶ **Understand Vocabulary** ••••••••••••••••••••••••

Use the review words to complete each sentence.

1. You can _____ to find a number that is close to the exact amount.

2. Numbers that are easy to compute with mentally are called

   _____.

3. The _____ is the amount left over when a number cannot be divided evenly.

4. A method of dividing in which multiples of the divisor are subtracted from the dividend and then the quotients are

   added together is called _____.

5. The number that is to be divided in a division problem is the

   _____.

6. The _____ is the number, not including the remainder, that results from dividing.

© Houghton Mifflin Harcourt Publishing Company

• **Interactive Student Edition**
• **Multimedia eGlossary**

Chapter 2 Vocabulary

compatible numbers

números compatibles

7

dividend

dividendo

18

divisor

divisor

19

factor

factor

27

partial quotient

cociente parcial

47

product

producto

54

quotient

cociente

57

remainder

residuo

59

The number that is to be divided in a division problem

Example: $36 \div 6$ or $6\overline{)36}$

dividend

Numbers that are easy to compute with mentally

A number multiplied by another number to find a product

Example: $46 \times 3 = 138$

factors

The number that divides the dividend

Example: $15 \div 3$ or $3\overline{)15}$

divisor

The answer to a multiplication problem

Example: $3 \times 15 = 45$

product

A method of dividing in which multiples of the divisor are subtracted from the dividend and then the quotients are added together

partial quotients

Example:

$$
\begin{array}{r}
5\overline{)125} \\
-50 \\
\hline 75 \\
-50 \\
\hline 25 \\
-25 \\
\hline 0
\end{array}
\qquad
\begin{array}{l}
10 \times 5 \\
\\
10 \times 5 \\
\\
5 \times 5 \\
\end{array}
\qquad
\begin{array}{r}
10 \\
\\
10 \\
\\
+5 \\
\hline 25
\end{array}
$$

The amount left over when a number cannot be divided equally

remainder

Example:

$$
\begin{array}{r}
102\,r2 \\
6\overline{)614} \\
-6 \\
\hline 01 \\
-0 \\
\hline 14 \\
-12 \\
\hline 2
\end{array}
$$

remainder

The number that results from dividing

Example: $8 \div 4 = 2$

quotient

Matchup

For 2–3 players

Materials

- 1 set of word cards

How to Play

1. Put the cards face-down in rows. Take turns to play.
2. Choose two cards and turn them over.
   - If the cards show a word and its meaning, it's a match. Keep the pair and take another turn.
   - If the cards do not match, turn them over again.
3. The game is over when all cards have been matched. The players count their pairs. The player with the most pairs wins.

Word Box

compatible numbers

dividend

divisor

factor

partial quotient

product

quotient

remainder

Journal

The Write Way

Reflect

Choose one idea. Write about it.

- Describe a situation in which you might use compatible numbers to estimate.
- Write a paragraph that uses at least **three** of these words.

  dividend   divisor   quotient   remainder

- Megan has $340 to spend on party favors for 16 guests. Tell how Megan can use partial quotients to figure out how much she can spend on each guest.
- A hiker wants to travel the same number of miles each day to complete a 128-mile trail. Explain and illustrate two different options for completing the trail. Draw your picture on another sheet of paper.

Place the First Digit

Essential Question How can you tell where to place the first digit of a quotient without dividing?

Common Core **Number and Operations in Base Ten—5.NBT.B.6**

MATHEMATICAL PRACTICES
MP1, MP4, MP6

Unlock the Problem

Tania has 8 purple daisies. In all, she counts 128 petals on her flowers. If each flower has the same number of petals, how many petals are on one flower?

- Underline the sentence that tells you what you are trying to find.
- Circle the numbers you need to use.
- How will you use these numbers to solve the problem?

_____

 Divide. 128 ÷ 8

STEP 1 Use an estimate to place the first digit in the quotient.

Estimate. 160 ÷ _____ = _____

The first digit of the quotient will be in

the _____ place.

STEP 2 Divide the tens.

$$\begin{array}{r} 1 \\ 8\overline{)128} \\ -\rule{1cm}{0.4pt} \\ \hline \rule{1cm}{0.4pt} \end{array}$$

Divide. 12 tens ÷ 8
Multiply. 8 × 1 ten

Subtract. 12 tens − _____ tens
Check. _____ tens cannot be shared among 8 groups without regrouping.

STEP 3 Regroup any tens left as ones. Then, divide the ones.

$$\begin{array}{r} 16 \\ 8\overline{)128} \\ -8\downarrow \\ \hline \\ - \\ \hline \end{array}$$

Divide. 48 ones ÷ 8
Multiply. 8 × 6 ones

Subtract. 48 ones − _____ ones
Check. _____ ones cannot be shared among 8 groups.

Since 16 is close to the estimate of _____ , the answer is reasonable.

So, there are 16 petals on one flower.

Math Talk MATHEMATICAL PRACTICES ⑥

Explain how estimating the quotient helps you at both the beginning and the end of a division problem.

Example

Divide. Use place value to place the first digit. 4,236 ÷ 5

STEP 1 Use place value to place the first digit.

$5\overline{)4,236}$

Look at the thousands.

4 thousands cannot be shared among 5 groups without regrouping.

Look at the hundreds.

_____ hundreds can be shared among 5 groups.

The first digit is in the _____ place.

STEP 2 Divide the hundreds.

$\dfrac{8}{5\overline{)4,236}}$

Divide. _____ hundreds ÷ _____

Multiply. _____ × _____ hundreds

Subtract. _____ hundreds − _____ hundreds

Check. _____ hundreds cannot be shared among 5 groups without regrouping.

STEP 3 Divide the tens.

$\begin{array}{r} 84 \\ 5\overline{)4,236} \\ -40\downarrow \\ \hline 23 \\ -20 \\ \hline 3 \end{array}$

Divide. _____

Multiply. _____

Subtract. _____

Check. _____

_____

STEP 4 Divide the ones.

$\begin{array}{r} 847 \\ 5\overline{)4,236} \\ -40 \\ \hline 23 \\ -20\downarrow \\ \hline 36 \\ -35 \\ \hline 1 \end{array}$

Divide. _____

Multiply. _____

Subtract. _____

Check. _____

_____

So, 4,236 ÷ 5 is _____ r_____.

 Math Talk

MATHEMATICAL PRACTICES ⑥

Explain how you know if your answer is reasonable.

Name _____

Divide.

1. 4)457

2. 5)1,035

3. 8)1,766

On Your Own

Math Talk

MATHEMATICAL PRACTICES 6

Use Math Vocabulary As you divide, explain how you know when to place a zero in the quotient.

Divide.

4. 8)275

5. 3)468

6. 4)3,220

7. 6)618

8. **GO DEEPER** Ryan earned $376 by working for 4 days. If he earned the same amount each day, how much could he earn working 5 days?

_____

Practice: Copy and Solve **Divide.**

9. 645 ÷ 8   10. 942 ÷ 6   11. 723 ÷ 7   12. 3,478 ÷ 9

13. 3,214 ÷ 5   14. 492 ÷ 4   15. 2,403 ÷ 9   16. 2,205 ÷ 6

17. **GO DEEPER** Will the first digit of the quotient of 2,589 ÷ 4 be in the hundreds or the thousands place? **Explain** how you can decide without finding the quotient.

_____

_____

_____

_____

🔑 Unlock the Problem Real World

18. **MATHEMATICAL PRACTICE ④ Interpret a Result** Rosa has a garden divided into sections. She has 125 daisy plants. If she plants an equal number of the daisy plants in each of 3 sections, how many daisy plants will be in each section? How many daisy plants will be left over?

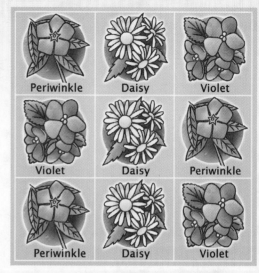

Periwinkle · Daisy · Violet
Violet · Daisy · Periwinkle
Periwinkle · Daisy · Violet

a. What information will you use to solve the problem?

_____

_____

b. How will you use division to find the number of daisy plants left over?

_____

_____

_____

c. Show the steps you use to solve the problem.
   Estimate: 120 ÷ 3 = _____

d. Complete the sentences:

   Rosa has _____ daisy plants. She puts an equal number in each

   of _____ sections.

   Each section has _____ plants. Rosa has _____ daisy plants left over.

19. **THINK SMARTER** One case can hold 3 boxes. Each box can hold 3 binders. How many cases are needed to hold 126 binders?

_____

20. **THINK SMARTER** For 20a–20b, choose Yes or No to indicate whether the first digit of the quotient is in the hundreds place.

20a. 1,523 ÷ 23   ○ Yes   ○ No

20b. 2,315 ÷ 9   ○ Yes   ○ No

Place the First Digit

Common Core   **COMMON CORE STANDARD—5.NBT.B.6**
*Perform operations with multi-digit whole
numbers and with decimals to hundredths.*

Divide.

1. 4)388

```
      97
   4)388
    −36
     28
    −28
      0
```

_____97_____

2. 3)579

_____

3. 8)712

_____

4. 9)204

_____

5. 2,117 ÷ 3

_____

6. 520 ÷ 8

_____

7. 1,812 ÷ 4

_____

8. 3,476 ÷ 6

_____

Problem Solving Real World

9. The school theater department made $2,142 on ticket sales for the three nights of their play. The department sold the same number of tickets each night and each ticket cost $7. How many tickets did the theater department sell each night?

_____

10. Andreus made $625 mowing yards. He worked for 5 consecutive days and earned the same amount of money each day. How much money did Andreus earn per day?

_____

11. **WRITE** ▸*Math* Write a word problem that must be solved by using division. Include the equation and the solution, and explain how to place the first digit in the quotient.

_____

_____

Lesson Check (5.NBT.B.6)

1. Kenny is packing cans into bags at the food bank. He can pack 8 cans into each bag. How many bags will Kenny need for 1,056 cans?

2. Liz polishes rings for a jeweler. She can polish 9 rings per hour. How many hours will it take her to polish 315 rings?

Spiral Review (5.NBT.A.2, 5.NBT.B.5, 5.NBT.B.6)

3. Fiona uses 256 fluid ounces of juice to make 1 bowl of punch. How many fluid ounces of juice will she use to make 3 bowls of punch?

4. Len wants to write the number 100,000 using a base of 10 and an exponent. What number should he use as the exponent?

5. A family pass to the amusement park costs $54. Using the Distributive Property, write an expression that can be used to find the cost in dollars of 8 family passes.

6. Gary is catering a picnic. There will be 118 guests at the picnic, and he wants each guest to have a 12-ounce serving of salad. How much salad should he make?

FOR MORE PRACTICE
GO TO THE
Personal Math Trainer

Name _____

Divide by 1-Digit Divisors

Essential Question How do you solve and check division problems?

Common Core **Number and Operations in Base Ten—5.NBT.B.6**

MATHEMATICAL PRACTICES
MP1, MP2, MP8

Unlock the Problem

Jenna's family is planning a trip to Oceanside, California. They will begin their trip in Scranton, Pennsylvania, and will travel 2,754 miles over 9 days. If the family travels an equal number of miles every day, how far will they travel each day?

- Underline the sentence that tells you what you are trying to find.
- Circle the numbers you need to use.

 Divide. 2,754 ÷ 9

STEP 1

Use an estimate to place the first digit in the quotient.

Estimate. 2,700 ÷ 9 = _____

The first digit of the quotient is in

the _____ place.

_____

STEP 2

Divide the hundreds.

_____

STEP 3

Divide the tens.

_____

STEP 4

Divide the ones.

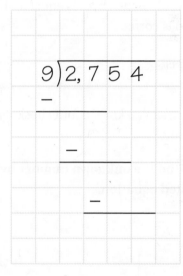

Since _____ is close to the estimate of _____, the answer is reasonable.

So, Jenna's family will travel _____ miles each day.

© Houghton Mifflin Harcourt Publishing Company

Math Talk MATHEMATICAL PRACTICES ❷

Reasoning Explain how you know the quotient is 306 and not 36.

CONNECT Division and multiplication are inverse operations. Inverse operations are opposite operations that undo each other. You can use multiplication to check your answer to a division problem.

 **Example** Divide. Check your answer.

To check your answer to a division problem, multiply the quotient by the divisor. If there is a remainder, add it to the product. The result should equal the dividend.

```
    102 r2
6)614
   -6
    01
   -0
    14
   -12
     2
```

```
  102   ← quotient
×   6   ← divisor
        
+   2   ← remainder
        ← dividend
```

Since the result of the check is equal to the dividend, the division is correct.

So, 614 ÷ 6 is _____.

You can use what you know about checking division to find an unknown value.

Try This! Find the unknown number by finding the value of *n* in the related equation.

A

```
      63
7)
```

$$n = 7 \times 63$$

↑ dividend   ↑ divisor   ↑ quotient

Multiply the divisor and the quotient.

$n =$ _____

B

```
    125 r
6)752
```

$$752 = 6 \times 125 + n$$

↑ dividend   ↑ divisor   ↑ quotient   ↖ remainder

Multiply the divisor and the quotient.

$$752 = 750 + n$$

Think: What number added to 750 equals 752?

$n =$ _____

Name _____

Divide. Check your answer.

1. 8)624   Check.

2. 4)3,220   Check.

3. 4)1,027   Check.

Math Talk

MATHEMATICAL PRACTICES ⑧

Generalize Explain how multiplication can help you check a quotient.

On Your Own

Divide.

4. 6)938

5. 4)762

6. 3)5,654

7. 8)475

Practice: Copy and Solve   Divide.

8. 4)671

9. 9)2,023

10. 3)4,685

11. 8)948

12. 1,326 ÷ 4

13. 5,868 ÷ 6

14. 566 ÷ 3

15. 3,283 ÷ 9

MATHEMATICAL PRACTICE ② Use Reasoning **Algebra** Find the value of n in each equation.
Write what n represents in the related division problem.

16. $n = 4 \times 58$

17. $589 = 7 \times 84 + n$

18. $n = 5 \times 67 + 3$

$n =$ _____

$n =$ _____

$n =$ _____

Problem Solving • Applications

Use the table to solve 19–21.

19. If the Welcome gold nugget were turned into 3 equal-sized gold bricks, how many troy ounces would each brick weigh?

_____

20. Pose a Problem Look back at Problem 19. Write a similar problem by changing the nugget and the number of bricks. Then solve the problem.

_____

_____

Large Gold Nuggets Found

Name	Weight	Location
Welcome Stranger	2,284 troy ounces	Australia
Welcome	2,217 troy ounces	Australia
Willard	788 troy ounces	California

WRITE ▸ *Math* • **Show Your Work** • • •

21. **GO DEEPER** Suppose the Willard gold nugget was turned into 4 equal-sized gold bricks. If one of the bricks was sold, how many troy ounces of the Willard nugget would be left?

_____

22. **THINK SMARTER** There are 246 students going on a field trip to pan for gold. If they are going in vans that hold 9 students each, how many vans are needed? How many students will ride in the van that isn't full?

23. **THINK SMARTER** Lily's teacher wrote the division problem on the board. Using the vocabulary box, label the parts of the division problem. Then, using the vocabulary, explain how Lily can check whether her teacher's quotient is correct.

_____

_____

quotient	divisor	dividend

$$\begin{array}{r} 82 \\ 9{\overline{)738}} \end{array}$$

Divide by 1-Digit Divisors

Common Core

COMMON CORE STANDARD—5.NBT.B.6
Perform operations with multi-digit whole numbers and with decimals to hundredths.

Divide.

1. 4)$\overline{724}$　　　**2.** 5)$\overline{312}$　　　**3.** $278 \div 2$　　　**4.** $336 \div 7$

```
    181
4)724
  -4
   32
  -32
   04
  - 4
    0
```

　　181 　　　_____　　_____　　_____

Find the value of *n* in each equation. Write what *n* represents in the related division problem.

5. $n = 3 \times 45$　　　　**6.** $643 = 4 \times 160 + n$　　　　**7.** $n = 6 \times 35 + 4$

_____　　　　_____　　　　_____

Problem Solving Real World

8. Randy has 128 ounces of dog food. He feeds his dog 8 ounces of food each day. How many days will the dog food last?

9. Angelina bought a 64-ounce can of lemonade mix. She uses 4 ounces of mix for each pitcher of lemonade. How many pitchers of lemonade can Angelina make from the can of mix?

_____　　　　　　　_____

10. **WRITE** ▸*Math* Use a map to plan a trip in the United States. Find the number of miles between your current location and your destination, and divide the mileage by the number of days or hours that you wish to travel.

_____

_____

Lesson Check (5.NBT.B.6)

1. A color printer will print 8 pages per minute. How many minutes will it take to print a report that has 136 pages?

2. A postcard collector has 1,230 postcards. If she displays them on pages that hold 6 cards each, how many pages does she need?

Spiral Review (5.NBT.A.1, 5.NBT.B.5, 5.NBT.B.6)

3. Francis is buying a stereo system for $196. She wants to pay for it in four equal monthly installments. What is the amount she will pay each month?

4. A bakery bakes 184 loaves of bread in 4 hours. How many loaves does the bakery bake in 1 hour?

5. Marvin collects trading cards. He stores them in boxes that hold 235 cards each. If Marvin has 4 boxes full of cards, how many cards does he have in his collection?

6. What is the value of the digit 7 in 870,541?

FOR MORE PRACTICE
GO TO THE
Personal Math Trainer

Division with 2-Digit Divisors

Essential Question How can you use base-ten blocks to model and understand division of whole numbers?

 Number and Operations in Base Ten—5.NBT.B.6
MATHEMATICAL PRACTICES
MP1, MP3, MP4, MP6

Investigate

Materials ■ base-ten blocks

There are 156 students in the Carville Middle School chorus. The music director wants the students to stand with 12 students in each row for the next concert. How many rows will there be?

A. Use base-ten blocks to model the dividend, 156.

B. Place 2 tens below the hundred to form a rectangle. How many groups of 12 does the rectangle show? How much of the dividend is not shown in this rectangle?

_____

C. Combine the remaining tens and ones into as many groups of 12 as possible. How many groups of 12 are there?

_____

D. Place these groups of 12 on the right side of the rectangle to make a larger rectangle.

E. The final rectangle shows _____ groups of 12.

So, there will be _____ rows of 12 students.

Draw Conclusions

1. **MATHEMATICAL PRACTICE ⑥** **Explain** why you still need to make groups of 12 after Step B.

_____

_____

2. **MATHEMATICAL PRACTICE ⑥** Describe how you can use base-ten blocks to **model** the quotient 176 ÷ 16.

_____

_____

_____

Make Connections

The two sets of groups of 12 that you found in the Investigate are partial quotients. First you found 10 groups of 12 and then you found 3 more groups of 12. Sometimes you may need to regroup before you can show a partial quotient.

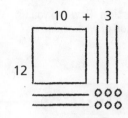

You can use a quick picture to record the partial products.

Divide. 180 ÷ 15

MODEL Use base-ten blocks.

STEP 1 Model the dividend, 180, as 1 hundred 8 tens.

Model the first partial quotient by making a rectangle with the hundred and 5 tens. In the Record section, cross out the hundred and tens you use.

The rectangle shows _____ groups of 15.

STEP 2 Additional groups of 15 cannot be made without regrouping.

Regroup 1 ten as 10 ones. In the Record section, cross out the regrouped ten.

There are now _____ tens and _____ ones.

STEP 3 Decide how many additional groups of 15 can be made with the remaining tens and ones. The number of groups is the second partial quotient.

Make your rectangle larger by including these groups of 15. In the Record section, cross out the tens and ones you use.

There are now _____ groups of 15.

So, 180 ÷ 15 is _____.

RECORD Use quick pictures.

Draw the first partial quotient.

Draw the first and second partial quotients.

MATHEMATICAL PRACTICES ④

Explain how your **model** shows the quotient.

Share and Show

Use the quick picture to divide.

1. 143 ÷ 13

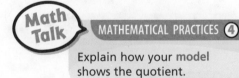

Name _____

Divide. Use base-ten blocks.

2. 168 ÷ 12

3. 154 ÷ 14

✓ **4.** 187 ÷ 11

Divide. Draw a quick picture.

5. 165 ÷ 11

6. 216 ÷ 18

✓ **7.** 182 ÷ 13

8. 228 ÷ 12

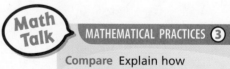

Math Talk

MATHEMATICAL PRACTICES ❸

Compare Explain how Exercise 7 is different from Exercises 6 and 8.

9. GO DEEPER On Monday, the Mars rover traveled 330 cm. On Tuesday, it traveled 180 cm. If the rover stopped every 15 cm to recharge, how many more times did it need to recharge on Monday than on Tuesday?

_____

Connect to Social Studies

Pony Express

The Pony Express used men riding horses to deliver mail between St. Joseph, Missouri, and Sacramento, California, from April, 1860 to October, 1861. The trail between the cities was approximately 2,000 miles long. The first trip from St. Joseph to Sacramento took 9 days 23 hours. The first trip from Sacramento to St. Joseph took 11 days 12 hours.

Solve.

10. **THINK SMARTER** Two Pony Express riders each rode part of a 176-mile trip. Each rider rode the same number of miles. They changed horses every 11 miles. How many horses did each rider use?

_____

11. **GO DEEPER** Suppose a Pony Express rider was paid $192 for 12 weeks of work. If he was paid the same amount each week, how much was he paid for 3 weeks of work?

_____

12. **MATHEMATICAL PRACTICE ①** **Analyze** Suppose three riders rode a total of 240 miles. If they used a total of 16 horses, and rode each horse the same number of miles, how many miles did they ride before replacing each horse?

_____

13. **THINK SMARTER** Suppose it took 19 riders a total of 11 days 21 hours to ride from St. Joseph to Sacramento. If they all rode the same number of hours, how many hours did each rider ride?

_____

14. **THINK SMARTER +** Scientists collect 144 rock samples. The samples will be divided among 12 teams of scientists for analysis. Draw a quick picture to show how the samples can be divided among the 12 teams.

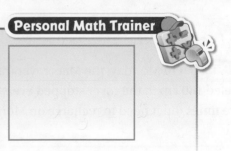

Personal Math Trainer

Division with 2-Digit Divisors

Common Core

COMMON CORE STANDARD—5.NBT.B.6
Perform operations with multi-digit whole numbers and with decimals to hundredths.

Use the quick picture to divide.

1. $132 \div 12 =$ ___11___

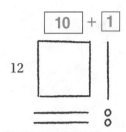

2. $168 \div 14 =$ _____

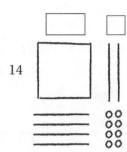

Divide. Draw a quick picture.

3. $192 \div 16 =$ _____

4. $169 \div 13 =$ _____

Problem Solving · Real World

5. There are 182 seats in a theater. The seats are evenly divided into 13 rows. How many seats are in each row?

6. There are 156 students at summer camp. The camp has 13 cabins. An equal number of students sleep in each cabin. How many students sleep in each cabin?

_____

_____

7. **WRITE** *Math* Write a division problem that has a 3-digit dividend and a divisor between 10 and 20. Show how to solve it by drawing a quick picture.

_____

_____

Lesson Check (5.NBT.B.6)

1. There are 198 students in the soccer league. There are 11 players on each soccer team. How many soccer teams are there?

2. Jason earned $187 for 17 hours of work. How much did Jason earn per hour?

Spiral Review (5.OA.A.2, 5.NBT.A.1, 5.NBT.B.5, 5.NBT.B.6)

3. What is the number written in standard form: six million, seven hundred thousand, twenty?

4. What is the following sentence written as an expression? "Add the product of 3 and 6 to 4."

5. To transport 228 people to an island, the island ferry makes 6 different trips. On each trip, the ferry carries the same number of people. How many people does the ferry transport on each trip?

6. Isabella sells 36 tickets to the school talent show. Each ticket costs $14. How much money does Isabella collect for the tickets she sells?

FOR MORE PRACTICE
GO TO THE
Personal Math Trainer

Name _____

Partial Quotients

Essential Question How can you use partial quotients to divide by 2-digit divisors?

Common Core

Number and Operations in Base Ten—5.NBT.B.6

MATHEMATICAL PRACTICES
MP1, MP3, MP8

Unlock the Problem

People in the United States eat about 23 pounds of pizza per person every year. If you ate that much pizza each year, how many years would it take you to eat 775 pounds of pizza?

- Rewrite in one sentence the problem you are asked to solve.

  _____

  _____

 Divide by using partial quotients.

775 ÷ 23

STEP 1

Subtract multiples of the divisor from the dividend until the remaining number is less than the multiple. The easiest partial quotients to use are multiples of 10.

STEP 2

Subtract smaller multiples of the divisor until the remaining number is less than the divisor. Then add the partial quotients to find the quotient.

COMPLETE THE DIVISION PROBLEM.

$$23\overline{)775}$$
$$-\ \_\_\_$$
$$545$$

10×23 | 10

775 ÷ 23 is _____ r _____.

So, it would take you more than 33 years to eat 775 pounds of pizza.

Remember

Depending on the question, a remainder may or may not be used in answering the question. Sometimes the quotient is adjusted based on the remainder.

🔓 Example

Myles is helping his father with the supply order for his pizza shop. For next week, the shop will need 1,450 ounces of mozzarella cheese. Each package of cheese weighs 32 ounces. Complete Myles's work to find how many packages of mozzarella cheese he needs to order.

```
32)1,450
  - 320      _____ × 32      [    ]
   1,130
   - 320      _____ × 32      [    ]
     810
    -320      _____ × 32      [    ]
     490
    -320      _____ × 32      [    ]
     170
    -160      _____ × 32    + [    ]
      10
```

1,450 ÷ 32 is _____ r _____.

So, he needs to order _____ packages of mozzarella cheese.

Math Talk

Generalize What does the remainder represent? Explain how a remainder will affect your answer.

Try This! Use different partial quotients to solve the problem above.

```
32)1,450
```

Math Idea

Using different multiples of the divisor to find partial quotients provides many ways to solve a division problem. Some ways are quicker, but all result in the same answer.

© Houghton Mifflin Harcourt Publishing Company • Image Credits: (tr) ©Steve Mason/Photodisc/Getty Images

Name _____

Divide. Use partial quotients.

1. $18\overline{)648}$

✓2. $62\overline{)3,186}$

✓3. $858 \div 57$

Math Talk

MATHEMATICAL PRACTICES ⑧

Generalize Explain what the greatest possible whole-number remainder is if you divide any number by 23.

On Your Own

Divide. Use partial quotients.

4. $73\overline{)584}$

5. $51\overline{)1,831}$

6. $82\overline{)2,964}$

7. $892 \div 26$

8. $1,056 \div 48$

9. $2,950 \div 67$

Practice: Copy and Solve Divide. Use partial quotients.

10. $653 \div 42$

11. $946 \div 78$

12. $412 \div 18$

13. $871 \div 87$

14. $1,544 \div 34$

15. $2,548 \div 52$

16. $2,740 \div 83$

17. $4,135 \div 66$

18. **GO DEEPER** The 5th grade is having a picnic this Friday. There will be 182 students and 274 adults. Each table seats 12 people. How many tables are needed?

Problem Solving · Applications Real World

Use the table to solve 19–22.

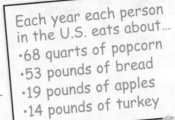

Each year each person
in the U.S. eats about…
•68 quarts of popcorn
•53 pounds of bread
•19 pounds of apples
•14 pounds of turkey

19. How many years would it take for a person in the United States to eat 855 pounds of apples?

_____

20. How many years would it take for a person in the United States to eat 1,120 pounds of turkey?

_____

21. **GO DEEPER** If 6 people in the United States each eat the average amount of popcorn for 5 years, how many quarts of popcorn will they eat?

_____

22. **MATHEMATICAL PRACTICE ❶ Make Sense of Problems** In the United States, a person eats more than 40,000 pounds of bread in a lifetime if he or she lives to be 80 years old. Does this statement make sense, or is it nonsense? Explain.

_____

_____

_____

23. **THINK SMARTER** In a study, 9 people ate a total of 1,566 pounds of potatoes in 2 years. If each person ate the same amount each year, how many pounds of potatoes did each person eat in 1 year?

_____

24. **THINK SMARTER** Nyree divided 495 by 24 using partial quotients. Find the quotient and remainder. Explain your answer using numbers and words.

$$24\overline{)495}$$

_____

_____

_____

Partial Quotients

Common Core

COMMON CORE STANDARD—5.NBT.B.6
*Perform operations with multi-digit whole
numbers and with decimals to hundredths.*

Divide. Use partial quotients.

1. 18)236

```
18)236
  −180   ← 10 × 18      10
───────
   56
  −36    ←  2 × 18       2
───────
   20
  −18    ←  1 × 18     + 1
───────
    2                   13
```

236 ÷ 18 is 13 r2.

2. 36)540

3. 27)624

4. 514 ÷ 28

5. 322 ÷ 14

6. 715 ÷ 25

Problem Solving Real World

7. A factory processes 1,560 ounces of olive oil per hour. The oil is packaged into 24-ounce bottles. How many bottles does the factory fill in one hour?

_____

8. A pond at a hotel holds 4,290 gallons of water. The groundskeeper drains the pond at a rate of 78 gallons of water per hour. How long will it take to drain the pond?

_____

9. **WRITE** ▸*Math* Explain how using partial quotients to divide is similar to using the Distributive Property to multiply.

_____

_____

_____

Lesson Check (5.NBT.B.6)

1. Yvette has 336 eggs to put into cartons. She puts one dozen eggs into each carton. How many cartons does she fill?

_____

2. Ned mows a 450 square-foot garden in 15 minutes. How many square feet of the garden does he mow in one minute?

_____

Spiral Review (5.NBT.A.1, 5.NBT.B.5, 5.NBT.B.6)

3. Raul has 56 bouncy balls. He puts the balls into 4 green gift bags. If he puts the same number of balls into each bag, how many balls does he put into each green bag?

_____

4. Marcia uses 5 ounces of chicken stock to make one batch of soup. She has a total of 400 ounces of chicken stock. How many batches of soup can Marcia make?

_____

5. Michelle buys 13 bags of gravel for her fish aquarium. If each bag weighs 12 pounds, how many pounds of gravel did she buy?

_____

6. What is the number 4,305,012 written in expanded notation?

_____

FOR MORE PRACTICE
GO TO THE
Personal Math Trainer

Name _____

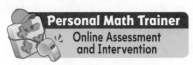

Personal Math Trainer
Online Assessment
and Intervention

Concepts and Skills

1. Explain how estimating the quotient helps you place the first
   digit in the quotient of a division problem. (5.NBT.B.6)

   _____

   _____

2. Explain how to use multiplication to check the answer to a
   division problem. (5.NBT.B.6)

   _____

   _____

   _____

Divide. (5.NBT.B.6)

3. 633 ÷ 3	**4.** 487 ÷ 8	**5.** 1,641 ÷ 4	**6.** 2,765 ÷ 9

Divide. Use partial quotients. (5.NBT.B.6)

7. 156 ÷ 13	**8.** 318 ÷ 53	**9.** 1,562 ÷ 34	**10.** 4,024 ÷ 68

11. Emma is planning a party for 128 guests. If 8 guests can be seated at each table, how many tables will be needed for seating at the party? (5.NBT.B.6)

_____

12. Tickets for the basketball game cost $14 each. If the sale of the tickets brought in $2,212, how many tickets were sold? (5.NBT.B.6)

_____

13. Margo used 864 beads to make necklaces for the art club. She made 24 necklaces with the beads. If each necklace has the same number of beads, how many beads did Margo use for each necklace? (5.NBT.B.6)

_____

14. Angie needs to buy 156 candles for a party. Each package has 8 candles. How many packages should Angie buy? (5.NBT.B.6)

_____

15. **GO DEEPER** Max delivers 8,520 pieces of mail in one year. If he delivers the same number of pieces of mail each month, about how many pieces of mail does he deliver in 2 months? Explain your steps. (5.NBT.B.6)

_____

_____

_____

Name _____

Estimate with 2-Digit Divisors

Essential Question How can you use compatible numbers to estimate quotients?

Common Core Number and Operations in Base Ten—5.NBT.B.6
MATHEMATICAL PRACTICES
MP1, MP2, MP3

CONNECT You can estimate quotients using compatible numbers that are found by using basic facts and patterns.

$$35 \div 5 = 7 \quad \leftarrow \text{basic fact}$$
$$350 \div 50 = 7$$
$$3{,}500 \div 50 = 70$$
$$35{,}000 \div 50 = 700$$

⚷ 🔓 Unlock the Problem

The observation deck of the Willis Tower in Chicago, Illinois, is 1,353 feet above the ground. Elevators lift visitors to that level in 60 seconds. About how many feet do the elevators travel per second?

◀ Willis Tower, formerly known as the Sears Tower, is the tallest building in the United States.

 Estimate. 1,353 ÷ 60

STEP 1

Use two sets of compatible numbers to find two different estimates.

1,353 ÷ 60	1,353 ÷ 60
↓	↓
1,200 ÷ 60	1,800 ÷ 60

STEP 2

Use patterns and basic facts to help estimate.

12 ÷ 6 = _____	18 ÷ 6 = _____
120 ÷ 60 = _____	_____ ÷ _____ = _____
1,200 ÷ 60 = _____	_____ ÷ _____ = _____

The elevators travel about _____ to _____ feet per second.

The more reasonable estimate is _____ because

_____ is closer to 1,353 than _____ is.

So, the observation deck elevators in the Willis Tower travel

about _____ feet per second.

🔒 Example Estimate money.

Miriam saved $650 to spend during her 18-day trip to Chicago. She doesn't want to run out of money before the trip is over, so she plans to spend about the same amount each day. Estimate how much she can spend each day.

Estimate. $18\overline{)\$650}$

$600 \div \underline{\hspace{1cm}} = \$30$ or $\underline{\hspace{2cm}} \div 20 = \$40$

So, Miriam can spend about \underline{\hspace{1cm}} to \underline{\hspace{1cm}} each day.

Math Talk

MATHEMATICAL PRACTICES ①

Analyze Would it be more reasonable to have an estimate or an exact answer for this example? Explain your reasoning.

- **MATHEMATICAL PRACTICE ②** **Use Reasoning** Which estimate do you think is the better one for Miriam to use? Explain your reasoning. _____

_____

_____

Try This! Use compatible numbers.

Find two estimates.	Estimate the quotient.
$52\overline{)415}$	$38\overline{)\$2,764}$

Share and Show MATH BOARD

Use compatible numbers to find two estimates.

1. $22\overline{)154}$

$140 \div 20 = \underline{\hspace{1cm}}$

$160 \div 20 = \underline{\hspace{1cm}}$

2. $68\overline{)503}$

3. $81\overline{)7,052}$

✓4. $33\overline{)291}$

✓5. $58\overline{)2,365}$

6. $19\overline{)5,312}$

114

© Houghton Mifflin Harcourt Publishing Company

On Your Own

Use compatible numbers to find two estimates.

7. $42 \overline{)396}$

8. $59 \overline{)413}$

9. $28 \overline{)232}$

Use compatible numbers to estimate the quotient.

10. $19 \overline{)228}$

11. $25 \overline{)\$595}$

12. $86 \overline{)7{,}130}$

13. GO DEEPER At an orchard, 486 green apples are to be organized into 12 green baskets and 633 red apples are to be organized into 31 red baskets. Use estimation to decide which color basket has more apples. About how many apples are in each basket of that color?

_____

14. A store owner bought a large box of 5,135 paper clips. He wants to repackage the paper clips into 18 smaller boxes. Each box should contain about the same number of paper clips. About how many paper clips should the store owner put into each box?

_____

15. Explain how you can use compatible numbers to estimate the quotient of $925 \div 29$.

_____

_____

_____

Problem Solving • Applications Real World

Use the picture to solve 16–17.

16. **THINK SMARTER** Use estimation to decide which building has the tallest floors. About how many meters is each floor?

_____

17. **MATHEMATICAL PRACTICE 3** Make Arguments About how many meters tall is each floor of the Chrysler Building? Use what you know about estimating quotients to justify your answer.

_____

_____

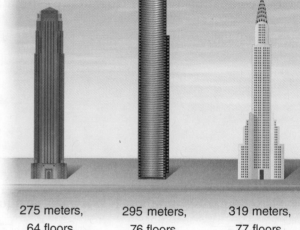

275 meters, 64 floors, Williams Tower, Texas

295 meters, 76 floors, Columbia Center, Washington

319 meters, 77 floors, Chrysler Building, New York

18. **WRITE** ▸Math Explain how you know whether the quotient of 298 ÷ 31 is closer to 9 or to 10.

_____

_____

_____

· · · · **WRITE** ▸Math · **Show Your Work** · · · ·

19. **GO DEEPER** Eli needs to save $235. To earn money, he plans to mow lawns and charge $21 for each. Write two estimates Eli could use to determine the number of lawns he needs to mow. Decide which estimate you think is the better one for Eli to use. Explain your reasoning.

_____

_____

_____

20. **THINK SMARTER** Anik built a tower of cubes. It was 594 millimeters tall. The height of each cube was 17 millimeters. About how many cubes did Anik use? Explain your answer.

_____

_____

Estimate with 2-Digit Divisors

Common Core · **COMMON CORE STANDARD—5.NBT.B.6**
Perform operations with multi-digit whole numbers and with decimals to hundredths.

Use compatible numbers to find two estimates.

1. $18\overline{)1,322}$

$1,200 \div 20 = 60$

$1,400 \div 20 = 70$

2. $12\overline{)478}$

3. $336 \div 12$

4. $2,242 \div 33$

Use compatible numbers to estimate the quotient.

5. $82\overline{)5,514}$

6. $61\overline{)5,320}$

7. $28\overline{)776}$

8. $23\overline{)1,624}$

Problem Solving (Real World)

9. A cubic yard of topsoil weighs 4,128 pounds. About how many 50-pound bags of topsoil can you fill with one cubic yard of topsoil?

_____

10. An electronics store places an order for 2,665 USB flash drives. One shipping box holds 36 flash drives. About how many boxes will it take to hold all the flash drives?

_____

11. **WRITE** *Math* Create a division problem with a 2-digit divisor. Using more than 1 set of compatible numbers, observe what happens when you estimate using a different divisor, a different dividend, and when both are different. Using a calculator, compare the estimates to the answer and describe the differences.

_____

_____

_____

_____

_____

_____

_____

Lesson Check (5.NBT.B.6)

1. Marcy has 567 earmuffs in stock. If she can put 18 earmuffs on each shelf, about how many shelves does she need for all the earmuffs?

2. Howard pays $327 for one dozen collector's edition baseball cards. About how much does he pay for each baseball card?

Spiral Review (5.NBT.A.1, 5.NBT.B.5, 5.NBT.B.6)

3. Andrew can frame 9 pictures each day. He has an order for 108 pictures. How many days will it take him to complete the order?

4. Madeleine can type 3 pages in one hour. How many hours will it take her to type a 123-page report?

5. Suppose you round 43,257,529 to 43,300,000. To what place value did you round the number?

6. Grace's catering company received an order for 118 apple pies. Grace uses 8 apples to make one apple pie. How many apples does she need to make all 118 pies?

© Houghton Mifflin Harcourt Publishing Company

FOR MORE PRACTICE
GO TO THE
Personal Math Trainer

Divide by 2-Digit Divisors

Essential Question How can you divide by 2-digit divisors?

Common Core Number and Operations in Base Ten—5.NBT.B.6
MATHEMATICAL PRACTICES
MP1, MP2, MP8

Unlock the Problem

Mr. Yates owns a smoothie shop. To mix a batch of his famous orange smoothies, he uses 18 ounces of freshly squeezed orange juice. Each day he squeezes 560 ounces of fresh orange juice. How many batches of orange smoothies can Mr. Yates make in a day?

- Underline the sentence that tells you what you are trying to find.
- Circle the numbers you need to use.

 Divide. 560 ÷ 18   **Estimate.** _____

STEP 1   Use the estimate to place the first digit in the quotient.

$18\overline{)560}$ The first digit of the quotient will be in the

_____ place.

STEP 2   Divide the tens.

$\begin{array}{r} 3 \\ 18\overline{)560} \\ -54 \\ \hline 2 \end{array}$

Divide. $56 \text{ tens} \div 18$

Multiply. _____

Subtract. _____

Check.   2 tens cannot be shared among 18 groups without regrouping.

STEP 3   Divide the ones.

$\begin{array}{r} 31 \text{ r2} \\ 18\overline{)560} \\ -54\downarrow \\ \hline 20 \\ -18 \\ \hline 2 \end{array}$

Divide. _____

Multiply. _____

Subtract. _____

Check. _____

Since 31 is close to the estimate of 30, the answer is reasonable.

So, Mr. Yates can make 31 batches of orange smoothies each day.

Math Talk   MATHEMATICAL PRACTICES ①

Describe what the remainder 2 represents.

🔑 Example

Every Wednesday, Mr. Yates orders fruit. He has set aside $1,250 to purchase Valencia oranges. Each box of Valencia oranges costs $41. How many boxes of Valencia oranges can Mr. Yates purchase?

You can use multiplication to check your answer.

Divide. 1,250 ÷ 41

DIVIDE	CHECK YOUR WORK

Estimate. _____

```
        30 r20
   41)1,250
    -
      ____
      ____
    -
      ____
      ____
```

```
      30
    × 41
    ─────
      30
  + 1,200
  ───────
             ____
        +    ____
           ───────
           1,250 ✓
```

So, Mr. Yates can buy _____ boxes of Valencia oranges.

Try This! **Divide. Check your answer.**

Ⓐ
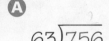
```
63)756
```

Ⓑ

```
22)4,692
```

Name _____

Divide. Check your answer.

1. 28)620

2. 64)842

3. 53)2,340

4. 723 ÷ 31

5. 1,359 ÷ 45

6. 7,925 ÷ 72

On Your Own

Math Talk

MATHEMATICAL PRACTICES 8

Generalize Explain why you can use multiplication to check division.

Divide. Check your answer.

7. 16)346

8. 34)421

9. 77)851

10. 21)1,098

11. 32)6,466

12. 45)9,500

13. **Go DEEPER** A city has 7,204 recycle bins. The city gives half of the recycle bins to its citizens. The rest of the recycle bins are divided into 23 equal groups for city parks. How many recycle bins are left over?

_____

Practice: Copy and Solve Divide. Check your answer.

14. 775 ÷ 35

15. 820 ÷ 41

16. 805 ÷ 24

17. 1,166 ÷ 53

18. 1,989 ÷ 15

19. 3,927 ÷ 35

Problem Solving · Applications

Use the list at the right to solve 20–22.

Smoothie Main Ingredients

Orange Tango Smoothie
18 ounces orange juice
12 ounces mango juice

Royal Purple Smoothie
22 ounces grape juice
8 ounces apple juice

Crazy Cranberry Smoothie
20 ounces cranberry juice
10 ounces passion fruit juice

20. **GO DEEPER** A smoothie shop receives a delivery of 968 ounces of grape juice and 720 ounces of orange juice. How many more Royal Purple smoothies than Orange Tango smoothies can be made with the shipment of juices?

21. **THINK SMARTER** The shop has 1,260 ounces of cranberry juice and 650 ounces of passion fruit juice. If the juices are used to make Crazy Cranberry smoothies, which juice will run out first? How much of the other juice will be left over?

22. **MATHEMATICAL PRACTICE ②** **Use Reasoning** In the refrigerator, there are 680 ounces of orange juice and 410 ounces of mango juice. How many Orange Tango smoothies can be made? Explain your reasoning.

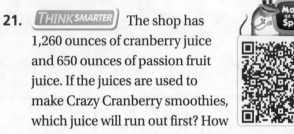

WRITE ▸Math · **Show Your Work**

Personal Math Trainer

23. **THINK SMARTER +** For 23a–23b, select True or False for each statement.

23a. 1,585 ÷ 16 is 99 r1. ○ True ○ False

23b. 1,473 ÷ 21 is 70 r7. ○ True ○ False

Divide by 2-Digit Divisors

Common Core   COMMON CORE STANDARD—5.NBT.B.6
*Perform operations with multi-digit whole
numbers and with decimals to hundredths.*

Divide. Check your answer.

1. 385 ÷ 12

$$\begin{array}{r} 32\ r1 \\ 12\overline{)385} \\ -36 \\ \hline 25 \\ -24 \\ \hline 1 \end{array}$$

2. 837 ÷ 36

3. 1,650 ÷ 55

4. 5,634 ÷ 18

5. 28)6,440

6. 52)5,256

7. 85)1,955

8. 46)5,624

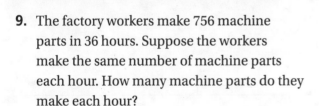

Problem Solving Real World

9. The factory workers make 756 machine parts in 36 hours. Suppose the workers make the same number of machine parts each hour. How many machine parts do they make each hour?

_____

10. One bag holds 12 bolts. Several bags filled with bolts are packed into a box and shipped to the factory. The box contains a total of 2,760 bolts. How many bags of bolts are in the box?

_____

11. **WRITE** ▸*Math* Choose a problem that you solved in the lesson, and solve the same problem using the partial quotients method. Compare the methods to solve the problems. Name the method you like better, and explain why.

_____

_____

_____

_____

Lesson Check (5.NBT.B.6)

1. A bakery packages 868 muffins into 31 boxes. The same number of muffins are put into each box. How many muffins are in each box?

2. Maggie orders 19 identical gift boxes. The Ship-Shape Packaging Company packs and ships the boxes for $1,292. How much does it cost to pack and ship each box?

Spiral Review (5.NBT.A.1, 5.NBT.B.6)

3. What is the standard form of the number four million, two hundred sixteen thousand, ninety?

4. Kelly and 23 friends go roller skating. They pay a total of $186. About how much does it cost for one person to skate?

5. In two days, Gretchen drinks seven 16-ounce bottles of water. She drinks the water in 4 equal servings. How many ounces of water does Gretchen drink in each serving?

6. What is the value of the underlined digit in 5,4<u>3</u>6,788?

© Houghton Mifflin Harcourt Publishing Company

FOR MORE PRACTICE
GO TO THE
Personal Math Trainer

Interpret the Remainder

Essential Question When solving a division problem, when do you write the remainder as a fraction?

 Number and Operations–Fractions—
5.NF.B.3 Also 5.NBT.B.6
MATHEMATICAL PRACTICES
MP2, MP3, MP4

Unlock the Problem

Scott and his family want to hike a trail that is 1,365 miles long. They will hike equal parts of the trail on 12 different hiking trips. How many miles will Scott's family hike on each trip?

- Circle the dividend you will use to solve the division problem.
- Underline the divisor you will use to solve the division problem.

When you solve a division problem with a remainder, the way you interpret the remainder depends on the situation and the question. Sometimes you need to use both the quotient and the remainder. You can do that by writing the remainder as a fraction.

One Way Write the remainder as a fraction.

First, divide to find the quotient and remainder.

Then, decide how to use the quotient and remainder to answer the question.

- The _____ represents the number of trips Scott and his family plan to take.

- The _____ represents the whole-number part of the number of miles Scott and his family will hike on each trip.

- The _____ represents the number of miles left over.

- The remainder represents 9 miles, which can also be divided into 12 parts and written as a fraction.

$$\frac{\text{remainder}}{\text{divisor}} \rightarrow \text{_____}$$

- Write the quotient with the remainder written as a fraction in simplest form.

_____

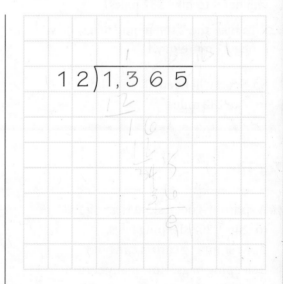

So, Scott and his family will hike _____ miles on each trip.

🔓 Another Way Use only the quotient.

The segment of the Appalachian Trail that runs through Pennsylvania is 232 miles long. Scott and his family want to hike 9 miles each day on the trail. How many days will they hike exactly 9 miles?

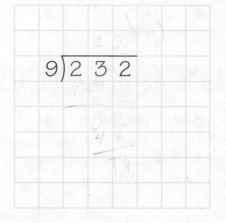

- Divide to find the quotient and the remainder.

- Since the remainder shows that there are not enough miles left for another 9-mile day, it is not used in the answer.

So, they will hike exactly 9 miles on each of _____ days.

🔓 Other Ways

A Add 1 to the quotient.

What is the total number of days that Scott will need to hike 232 miles?

- To hike the 7 remaining miles, he will need 1 more day.

So, Scott will need _____ days to hike 232 miles.

B Use the remainder as the answer.

If Scott hikes 9 miles each day except the last day, how many miles will he hike on the last day?

- The remainder is 7.

So, Scott will hike _____ miles on the last day.

Try This!

A sporting goods store is going to ship 1,252 sleeping bags. Each shipping carton can hold 8 sleeping bags. How many cartons are needed to ship all of the sleeping bags?

$$
\begin{array}{r}
1 \\
8\overline{)1{,}252} \\
-8 \\
\hline
45 \\
- \\
\hline
2 \\
- \\
\hline
4
\end{array}
$$

Since there are _____ sleeping bags left over,

_____ cartons will be needed for all of the sleeping bags.

Math Talk

MATHEMATICAL PRACTICES ④

Modeling Explain why you would not write the remainder as a fraction when you find the number of cartons needed in the Try This.

Name _____

Interpret the remainder to solve.

1. Erika and Bradley want to hike the Big Cypress Trail. They will hike a total of 75 miles. If Erika and Bradley plan to hike for 12 days, how many miles will they hike each day?

   a. Divide to find the quotient and remainder.

   b. Decide how to use the quotient and remainder to answer the question.

   _____

$$1\,2\,\overline{)7\,5}\quad r$$

2. **What if** Erika and Bradley want to hike 14 miles each day? How many days will they hike exactly 14 miles?

   _____

3. Dylan's hiking club is planning to stay overnight at a camping lodge. Each large room can hold 15 hikers. There are 154 hikers. How many rooms will they need?

   _____

On Your Own

Interpret the remainder to solve.

4. **GO DEEPER** The students in a class of 24 share 48 apple slices and 36 orange slices equally among them. How many pieces of fruit did each student get?

   _____

5. Fiona has 212 stickers to put in her sticker book. Each page holds 18 stickers. How many pages does Fiona need for all of her stickers?

   _____

6. A total of 123 fifth-grade students are going to Fort Verde State Historic Park. Each bus holds 38 students. All of the buses are full except one. How many students will be in the bus that is not full?

   _____

7. **MATHEMATICAL PRACTICE ③ Verify the Reasoning of Others** Sheila is going to divide a 36-inch piece of ribbon into 5 equal pieces. She says each piece will be 7 inches long. What is Sheila's error?

   _____
   _____
   _____

🔑 Unlock the Problem

8. Maureen has 243 ounces of trail mix. She puts an equal number of ounces in each of 15 bags. How many ounces of trail mix does Maureen have left over?

a. What do you need to find? _____

b. How will you use division to find how many ounces of trail mix are left over?

_____

_____

c. Show the steps you use to solve the problem.

d. Complete the sentences.

Maureen has _____ ounces of trail mix.

She puts an equal number of ounces in each

of _____ bags.

Each bag has _____ ounces.

Maureen has _____ ounces of
trail mix left over.

9. THINK SMARTER James has 884 feet of rope. There are 12 teams of hikers. If James gives an equal amount of rope to each team, how much rope will each team receive?

_____

10. THINK SMARTER Rory works at a produce packing plant. She packed 2,172 strawberries last week and put them in containers with 8 strawberries in each one. How many containers of strawberries did Rory fill with 8 strawberries? Explain how you used the quotient and the remainder to answer the question.

_____

_____

_____

Interpret the Remainder

Common Core

COMMON CORE STANDARD—5.NF.B.3
Apply and extend previous understandings of multiplication and division to multiply and divide fractions.

Interpret the remainder to solve.

1. Warren spent 140 hours making 16 wooden toy trucks for a craft fair. If he spent the same amount of time making each truck, how many hours did he spend making each truck?

$$
\begin{array}{r}
8 \\
16 \overline{)140} \\
-128 \\
\hline
12
\end{array}
$$

$8\frac{3}{4}$ hours

2. Marcia has 412 flowers for centerpieces. She uses 8 flowers for each centerpiece. How many centerpieces can she make?

Problem Solving · Real World

3. A campground has cabins that can each hold 28 campers. There are 148 campers visiting the campground. How many cabins are full if 28 campers are in each cabin?

4. Jenny has 220 ounces of cleaning solution that she wants to divide equally among 12 large containers. How much cleaning solution should she put in each container?

5. **WRITE** *Math* Suppose you have 192 marbles in groups of 15 marbles each. Find the number of groups of marbles that you have. Write the quotient with the remainder written as a fraction. Explain what the fraction part of your answer means.

_____

_____

_____

_____

Lesson Check (5.NF.B.3)

1. Henry and 28 classmates go to the roller skating rink. Each van can hold 11 students. If all of the vans are full except one, how many students are in the van that is not full?

_____

2. Candy buys 20 ounces of mixed nuts. She puts an equal number of ounces in each of 3 bags. How many ounces of mixed nuts will be in each bag? Write the answer as a whole number and a fraction.

_____

Spiral Review (5.NBT.B.5, 5.NBT.B.6)

3. Jayson earns $196 each week bagging groceries at the store. He saves half his earnings each week. How much money does Jayson save per week?

_____

4. Desiree swims laps for 25 minutes each day. How many minutes does she spend swimming laps in 14 days?

_____

5. Steve is participating in a bike-a-thon for charity. He will bike 144 miles per day for 5 days. How many miles will Steve bike in the five days?

_____

6. Kasi is building a patio. He has 136 bricks. He wants the patio to have 8 rows, each with the same number of bricks. How many bricks will Kasi put in each row?

_____

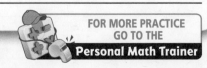

FOR MORE PRACTICE
GO TO THE
Personal Math Trainer

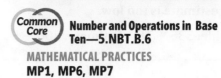
Adjust Quotients

Essential Question How can you adjust the quotient if your estimate is too high or too low?

Common Core
Number and Operations in Base Ten—5.NBT.B.6
MATHEMATICAL PRACTICES
MP1, MP6, MP7

CONNECT When you estimate to decide where to place the first digit, you can also try using the first digit of your estimate to find the first digit of your quotient. Sometimes an estimate is too low or too high.

Divide. 3,382 ÷ 48

Estimate. 3,000 ÷ 50 = 60

Try 6 tens.

If an estimate is too low, the difference will be greater than the divisor.

$$\begin{array}{r} 6 \\ 48\overline{)3{,}382} \\ -2\,88 \\ \hline 50 \end{array}$$

Since the estimate is too low, adjust by increasing the number in the quotient.

Divide. 453 ÷ 65

Estimate. 490 ÷ 70 = 7

Try 7 ones.

If an estimate is too high, the product with the first digit will be too large and cannot be subtracted.

$$\begin{array}{r} 7 \\ 65\overline{)453} \\ -455 \end{array}$$

Since the estimate is too high, adjust by decreasing the number in the quotient.

Unlock the Problem Real World

A new music group makes 6,127 copies of its first CD. The group sells 75 copies of the CD at each of its shows. How many shows does it take the group to sell all of the CDs?

 Divide. 6,127 ÷ 75   **Estimate.** 6,300 ÷ 70 = 90

STEP 1 Use the estimate, 90. Try 9 tens.

• Is the estimate too high, too low, or correct?

_____

• Adjust the number in the quotient if needed.

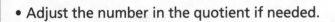

STEP 2 Estimate the next digit in the quotient.
      Divide the ones.
      Estimate: 140 ÷ 70 = 2. Try 2 ones.

• Is the estimate too high, too low, or correct?

_____

• Adjust the number in the quotient if needed.

$$75\overline{)6{,}127}$$

So, it takes the group _____ shows to sell all of the CDs.

Try This! When the difference is equal to or greater than the divisor, the estimate is too low.

Divide. 336 ÷ 48   **Estimate.** 300 ÷ 50 = 6

Use the estimate.

Try 6 ones.

$$48\overline{)336}^{6}$$

Since _____, the estimate is _____.

336 ÷ 48 = _____

Adjust the estimated digit in the quotient if needed. Then divide.

Try _____.

Math Talk
MATHEMATICAL PRACTICES ⑥

Explain why using the closest estimate could be useful in solving a division problem.

Share and Show MATH BOARD

Adjust the estimated digit in the quotient, if needed. Then divide.

1. $41\overline{)1{,}546}^{4}$

2. $16\overline{)416}^{2}$

✓ 3. $34\overline{)2{,}831}^{9}$

Divide.

4. $19\overline{)915}$

5. $28\overline{)1{,}825}$

✓ 6. $45\overline{)3{,}518}$

Math Talk
MATHEMATICAL PRACTICES ①

Evaluate Explain how you know whether an estimated quotient is too low or too high.

Name _____

On Your Own

Divide.

7. 15$\overline{)975}$

8. 37$\overline{)264}$

9. 34$\overline{)6,837}$

Practice: Copy and Solve Divide.

10. $452 \div 31$

11. $592 \div 74$

12. $785 \div 14$

13. $601 \div 66$

14. $1,067 \div 97$

15. $2,693 \div 56$

16. $1,488 \div 78$

17. $2,230 \div 42$

18. $4,295 \div 66$

MATHEMATICAL PRACTICE ⑦ Identify Relationships **Algebra** Write the unknown number for each ■.

19. ■ $\div 33 = 11$

20. $1,092 \div 52 = $ ■

21. $429 \div $ ■ $= 33$

■ = _____

■ = _____

■ = _____

22. **MATHEMATICAL PRACTICE ⑥** Explain a Method A deli served 1,288 sandwiches in 4 weeks. If it served the same number of sandwiches each day, how many sandwiches did it serve in 1 day? Explain how you found your answer.

_____

_____

_____

23. **THINK SMARTER** Kainoa collects trading cards. He has 1,025 baseball cards, 713 basketball cards, and 836 football cards. He wants to put all of them in albums. Each page in the albums holds 18 cards. How many pages will he need to hold all of his cards?

_____

Unlock the Problem

24. **GO DEEPER** A banquet hall serves 2,394 pounds of turkey during a 3-week period. If the same amount is served each day, how many pounds of turkey does the banquet hall serve each day?

a. What do you need to find? _____

_____

b. What information are you given? _____

_____

c. What other information will you use?

_____

_____

_____

d. Find how many days there are in 3 weeks.

There are _____ days in 3 weeks.

e. Divide to solve the problem.

f. Complete the sentence.

The banquet hall serves _____ of turkey each day.

25. Marcos mixes 624 ounces of lemonade. He wants to fill the 52 cups he has with equal amounts of lemonade. How much lemonade should he put in each cup?

_____

26. **THINK SMARTER** Oliver estimates the first digit in the quotient.

$$\begin{array}{r} 9 \\ 75\overline{)6,234} \end{array}$$

Oliver's estimate is

| correct. |
| too high. |
| too low |

Adjust Quotients

Adjust the estimated digit in the quotient, if needed. Then divide.

COMMON CORE STANDARD—5.NBT.B.6
Perform operations with multi-digit whole numbers and with decimals to hundredths.

1.

$$
\begin{array}{r}
5 \\
16\overline{)976} \\
-80 \\
\hline
17
\end{array}
$$

$$
\begin{array}{r}
61 \\
16\,\overline{)976} \\
-96 \\
\hline
16 \\
-16 \\
\hline
0
\end{array}
$$

2.

$$
\begin{array}{r}
3 \\
24\overline{)689}
\end{array}
$$

3.

$$
\begin{array}{r}
3 \\
65\overline{)2,210}
\end{array}
$$

4.

$$
\begin{array}{r}
2 \\
38\overline{)7,035}
\end{array}
$$

Divide.

5. $2,961 \div 47$

6. $2,072 \div 86$

7. $44\overline{)2,910}$

8. $82\overline{)4,018}$

Problem Solving · Real World

9. A copier prints 89 copies in one minute. How long does it take the copier to print 1,958 copies?

_____

10. Erica is saving her money to buy a dining room set that costs $580. If she saves $29 each month, how many months will she need to save to have enough money to buy the set?

_____

11. **WRITE** ▸*Math* Explain the different ways that you can use multiplication to estimate and solve division problems.

_____

_____

_____

_____

_____

Lesson Check (5.NBT.B.6)

1. Gail ordered 5,675 pounds of flour for the bakery. The flour comes in 25-pound bags. How many bags of flour will the bakery receive?

2. Simone is in a bike-a-thon for a fundraiser. She receives $15 in pledges for every mile she bikes. If she wants to raise $510, how many miles does she need to bike?

Spiral Review (5.OA.A.2, 5.NBT.A.1, 5.NBT.B.6)

3. Lina makes beaded bracelets. She uses 9 beads to make each bracelet. How many bracelets can she make with 156 beads?

4. A total of 1,056 students from different schools enter the county science fair. Each school enters exactly 32 students. How many schools participate in the science fair?

5. What is $\frac{1}{10}$ of 6,000?

6. Christy buys 48 barrettes. She shares the barrettes equally between herself and her 3 sisters. Write an expression to represent the number of barrettes each girl gets.

FOR MORE PRACTICE
GO TO THE
Personal Math Trainer

Name _____

Problem Solving • Division

Essential Question How can the strategy *draw a diagram* help you solve a division problem?

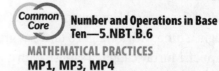

Common Core Number and Operations in Base Ten—5.NBT.B.6

MATHEMATICAL PRACTICES
MP1, MP3, MP4

 Unlock the Problem Real World

Sean and his family chartered a fishing boat for the day. Sean caught a blue marlin and an amberjack. The weight of the blue marlin was 12 times as great as the weight of the amberjack. The combined weight of both fish was 273 pounds. How much did each fish weigh?

Read the Problem

What do I need to find?	**What information do I need to use?**	**How will I use the information?**
I need to find _____ _____.	I need to know that Sean caught a total of _____ pounds of fish and the weight of the blue marlin was _____ times as great as the weight of the amberjack.	I can use the strategy _____ and then divide. I can draw and use a bar model to write the division problem that helps me find the weight of each fish.

Solve the Problem

I will draw one box to show the weight of the amberjack. Then I will draw a bar of 12 boxes of the same size to show the weight of the blue marlin. I can divide the total weight of the two fish by the total number of boxes.

amberjack [ ]

blue marlin [ ][ ][ ][ ][ ][ ][ ][ ][ ][ ][ ][ ] } 273 pounds

$$\begin{array}{r} 2 \\ 13\overline{)273} \\ -26 \\ \hline \end{array}$$

Write the quotient in each box. Multiply it by 12 to find the weight of the blue marlin.

So, the amberjack weighed _____ pounds and the

blue marlin weighed _____ pounds.

🔒 Try Another Problem

Jason, Murray, and Dana went fishing. Dana caught a red snapper. Jason caught a tuna with a weight 3 times as great as the weight of the red snapper. Murray caught a sailfish with a weight 12 times as great as the weight of the red snapper. If the combined weight of the three fish was 208 pounds, how much did the tuna weigh?

Read the Problem

What do I need to find?	What information do I need to use?	How will I use the information?

Solve the Problem

So, the tuna weighed _____ pounds.

- How can you check if your answer is correct? _____

_____

MATHEMATICAL PRACTICES ①

Analyze Explain how you could use another strategy to solve this problem.

Name _____

1. Paula caught a tarpon with a weight that was 10 times as great as the weight of a permit fish she caught. The total weight of the two fish was 132 pounds. How much did each fish weigh?

    First, draw one box to represent the weight of the permit fish and ten boxes to represent the weight of the tarpon.

    Next, divide the total weight of the two fish by the total number of boxes you drew. Place the quotient in each box.

    Last, find the weight of each fish.

    The permit fish weighed _____ pounds.

    The tarpon weighed _____ pounds.

· · · · · · · · **WRITE** ▸ *Math* · **Show Your Work** · · · · · ·

2. What if the weight of the tarpon was 11 times the weight of the permit fish, and the total weight of the two fish was 132 pounds? How much would each fish weigh?

    permit fish: _____ pounds

    tarpon: _____ pounds

3. Jon caught four fish that weighed a total of 252 pounds. The kingfish weighed twice as much as the amberjack and the white marlin weighed twice as much as the kingfish. The weight of the tarpon was 5 times the weight of the amberjack. How much did each fish weigh?

    amberjack: _____ pounds

    kingfish: _____ pounds

    marlin: _____ pounds

    tarpon: _____ pounds

On Your Own

Use the table to solve 4–5.

4. [THINK SMARTER] Kevin bought 3 bags of gravel to cover the bottom of his fish tank. He has 8 pounds of gravel left over. How much gravel did Kevin use to cover the bottom of the tank?

_____

5. [MATHEMATICAL PRACTICE ③] **Apply** Look back at Problem 4. Write a similar problem by changing the number of bags of gravel and the amount of gravel left.

_____

_____

_____

Kevin's Supply List for a Saltwater Aquarium	
40-gal tank	$170
Aquarium light	$30
Filtration system	$65
Thermometer	$2
15-lb bag of gravel	$13
Large rocks	$3 per lb
Clown fish	$20 each
Damselfish	$7 each

6. [THINK SMARTER] The crew on a fishing boat caught four fish that weighed a total of 1,092 pounds. The tarpon weighed twice as much as the amberjack and the white marlin weighed twice as much as the tarpon. The weight of the tuna was 5 times the weight of the amberjack. How much did each fish weigh?

_____

_____

7. [GO DEEPER] A fish market bought two swordfish at a rate of $13 per pound. The cost of the larger fish was 3 times as great as the cost of the smaller fish. The total cost of the two fish was $3,952. How much did each fish weigh?

_____

_____

Personal Math Trainer

8. [THINK SMARTER +] Eric and Stephanie took their younger sister Melissa to pick apples. Eric picked 4 times as many apples as Melissa. Stephanie picked 6 times as many apples as Melissa. Eric and Stephanie picked 150 apples together. Draw a diagram to find the number of apples Melissa picked.

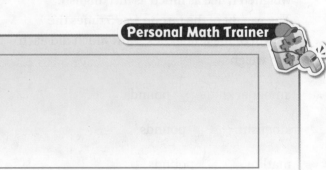

Problem Solving • Division

 COMMON CORE STANDARD—5.NBT.B.6
Perform operations with multi-digit whole numbers and with decimals to hundredths.

Show your work. Solve each problem.

1. Duane has 12 times as many baseball cards as Tony. Between them, they have 208 baseball cards. How many baseball cards does each boy have?

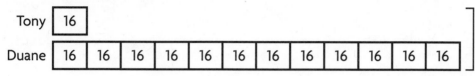

$208 \div 13 = 16$

Tony: 16 cards; Duane: 192 cards

2. Hallie has 10 times as many pages to read for her homework assignment as Janet. Altogether, they have to read 264 pages. How many pages does each girl have to read?

3. Kelly has 4 times as many songs on her music player as Lou. Tiffany has 6 times as many songs on her music player as Lou. Altogether, they have 682 songs on their music players. How many songs does Kelly have?

4. **WRITE** ▸ *Math* Create a word problem that uses division. Draw a bar model to help you write an equation to solve the problem.

Lesson Check (5.NBT.B.6)

1. Chelsea has 11 times as many art brushes as Monique. If they have 60 art brushes altogether, how many brushes does Chelsea have?

2. Jo has a gerbil and a German shepherd. The shepherd eats 14 times as much food as the gerbil. Altogether, they eat 225 ounces of dry food per week. How many ounces of food does the German shepherd eat per week?

Spiral Review (5.NBT.B.5, 5.NBT.B.6, 5.NF.B.3)

3. Jeanine is twice as old as her brother Marc. If the sum of their ages is 24, how old is Jeanine?

4. Larry is shipping nails that weigh a total of 53 pounds. He divides the nails equally among 4 shipping boxes. How many pounds of nails does he put in each box?

5. Annie plants 6 rows of small flower bulbs in a garden. She plants 132 bulbs in each row. How many bulbs does Annie plant?

6. Next year, four elementary schools will each send 126 students to Bedford Middle School. What is the total number of students the elementary schools will send to the middle school?

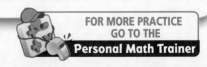

FOR MORE PRACTICE
GO TO THE
Personal Math Trainer

Name ______________________________

Chapter 2 Review/Test

Personal Math Trainer
Online Assessment and Intervention

1. Choose the word that makes the sentence true.
The first digit in the quotient of 1,875 ÷ 9

will be in the [ones / tens / hundreds / thousands] place.

2. For 2a–2d, select True or False to indicate whether the quotient is correct.

2a. 225 ÷ 9 = 25	○ True	○ False
2b. 154 ÷ 7 = 22	○ True	○ False
2c. 312 ÷ 9 = 39	○ True	○ False
2d. 412 ÷ 2 = 260	○ True	○ False

3. Chen is checking a division problem by doing the following:

$$\begin{array}{r} 152 \\ \times \quad 4 \\ \hline \square \\ + \quad 2 \\ \hline \square \end{array}$$

What problem is Chen checking?

GO DIGITAL Assessment Options Chapter Test

4. Isaiah wrote this problem in his notebook. Using the vocabulary boxes, label the parts of the division problem. Then, using the vocabulary, explain how Isaiah can check whether his quotient is correct.

quotient	divisor	dividend

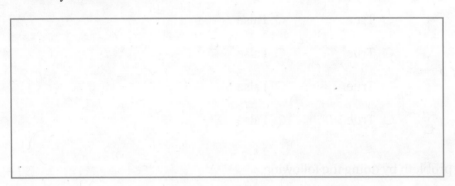

$$72 \quad \boxed{}$$
$$\boxed{} \; 9\overline{)648} \quad \boxed{}$$

_____

_____

_____

5. Tammy says the quotient of 793 ÷ 6 is 132 r1. Use multiplication to show if Tammy's answer is correct.

6. Jeffery wants to save the same amount of money each week to buy a new bike. He needs $252. If he wants the bike in 14 weeks, how much money should Jeffery save each week?

$ _____

7. Dana is making a seating chart for an awards banquet. There are 184 people coming to the banquet. If 8 people can be seated at each table, how many tables will be needed for the awards banquet?

_____ tables

Name _____

8. Darrel divided 575 by 14 by using partial quotients. What is the quotient? Explain your answer using numbers and words.

$$14\overline{)575}$$
$$-\quad 10 \times 14 \quad\vdots\quad 10$$
$$435$$

_____

_____

_____

9. For 9a–9c, choose Yes or No to indicate whether the statement is correct.

9a. $5{,}210 \div 17$ is 306 r8. ○ Yes   ○ No

9b. $8{,}808 \div 42$ is 209 r30. ○ Yes   ○ No

9c. $1{,}248 \div 24$ is 51. ○ Yes   ○ No

10. Divide. Draw a quick picture.

$156 \div 12 =$ ☐

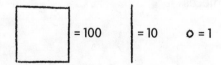

☐ = 100 | = 10   o = 1

11. Divide. Show your work.

$17\overline{)5,210}$

12. Choose the compatible numbers that will give the best estimate for $429 \div 36$.

○ 300
○ 350   and
○ 440

○ 60
○ 50
○ 40

13. **GO DEEPER** Samuel needs 233 feet of wood to build a fence. The wood comes in lengths of 11 feet.

Part A

How many total pieces of wood will Samuel need? Explain your answer.

Part B

Theresa needs twice as many feet of wood as Samuel. How many pieces of wood does Theresa need? Explain your answer.

Personal Math Trainer

14. **THINK SMARTER +** Russ and Vickie are trying to solve this problem:
There are 146 students taking buses to the museum. If each
bus holds 24 students, how many buses will they need?

Russ says the students need 6 buses. Vickie says they need 7 buses.
Who is correct? Explain your reasoning.

15. Write the letter for each quick picture under the division problem it
represents.

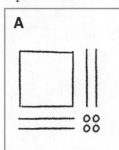
A

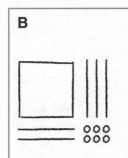
B

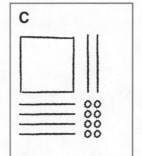
C

| $156 \div 12 = 13$ | $168 \div 12 = 14$ | $144 \div 12 = 12$ |

16. Steve is buying apples for the fifth grade. Each bag holds 12 apples. If there are 75 students total, how many bags of apples will Steve need to buy if he wants to give one apple to each student?

_____ bags

17. Rasheed needs to save $231. To earn money, he plans to wash cars and charge $12 per car. Write two estimates Rasheed could use to determine how many cars he needs to wash.

18. Paula has a dog that weighs 3 times as much as Carla's dog. The total weight of the dogs is 48 pounds. How much does Paula's dog weigh?

Draw a diagram to find the weight of Paula's dog.

19. Dylan estimates the first digit in the quotient.

Dylan's estimate is

| too high. |
| too low |

.